THE CASE FOR KILLER ROBOTS

The Case for Killer Robots

Why America's Military Needs to Continue Development of Lethal AI

ROBERT J. MARKS

SEATTLE   DISCOVERY INSTITUTE PRESS   2020

Description

Doomsday headlines warn that the age of "killer robots" is upon us, and that new military technologies based on artificial intelligence (AI) will lead to the annihilation of the human race. What's fact and what's fiction? In *The Case for Killer Robots: Why America's Military Needs to Continue Development of Lethal AI*, artificial intelligence expert Robert J. Marks investigates the potential military use of lethal AI and examines the practical and ethical challenges. Marks provocatively argues that the development of lethal AI is not only appropriate in today's society—it is unavoidable if America wants to survive and thrive into the future. Dr. Marks directs the Walter Bradley Center for Natural and Artificial Intelligence at Discovery Institute, and he is a Distinguished Professor of Electrical and Computer Engineering at Baylor University. This short monograph is produced in conjunction with the Walter Bradley Center for Natural and Artificial Intelligence, which can be visited at centerforintelligence.org.

Library Cataloging Data

The Case for Killer Robots: Why America's Military Needs to Continue Development of Lethal AI by Robert J. Marks

64 pages, 6 x 9 x 0.13 inches & 0.22 lb, 229 x 152 x 3 mm. & 0.1 kg

Library of Congress Control Number: 2019955448

ISBN-13: 978-1-936599-77-6 (paperback), 978-1-936599-79-0 (Kindle), 978-1-936599-79-0 (EPUB)

BISAC:

POL069000 POLITICAL SCIENCE / Public Policy / Military Policy

TEC025000 TECHNOLOGY & ENGINEERING / Military Science

COM014000 COMPUTERS / Computer Science

POL063000 POLITICAL SCIENCE / Public Policy / Science & Technology Policy

Publisher Information

Discovery Institute Press, 208 Columbia Street, Seattle, WA 98104

Internet: discoveryinstitutepress.com

Published in the United States of Ameria on acid-free paper.

First edition, first printing, January 2020.

ENDORSEMENTS

"This book is a succinct, well-reasoned, detailed and provocative voice in one of the most important conversations of our time. It should be read by anyone with an interest in the moral and social implications of AI."

—Donald C. Wunsch II, PhD, Mary K. Finley Missouri Distinguished Professor of Computer Engineering, Missouri University of Science and Technology; Director, Applied Computational Intelligence Laboratory, Missouri University of Science and Technology

"Science fiction-fed fears of killer robots 'waking up' and taking over the world prevent us from facing this basic fact: Bad guys have a say in what the world is like. A decision not to develop AI for defense is a choice to allow our most vicious enemies to develop superior weaponry to threaten and kill the innocent. It also means our weapons will be more rather than less likely to harm and kill non-combatants. In this vitally important book, Robert Marks makes a lucid and compelling case that we have a moral obligation to develop lethal AI. He also reminds us that moral questions apply, not to the tools that we use to protect ourself, but to how we use them when war becomes a necessity."

—Jay Richards, PhD, Research Assistant Professor, Busch School of Business, The Catholic University of America; author, *The Human Advantage: The Future of American Work in an Age of Smart Machines*

CONTENTS

THE BRADLEY CENTER

*T*HE CASE FOR *KILLER ROBOTS* IS A POSITION PAPER OF THE WAL-
ter Bradley Center for Natural and Artificial Intelligence at Discovery Institute. The mission of the Bradley Center is to "explore the
benefits as well as the challenges raised by artificial intelligence (AI) in
light of the enduring truth of human exceptionalism."[1]

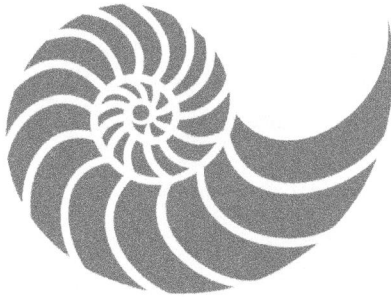

The *Walter Bradley* Center for
NATURAL & ARTIFICIAL
INTELLIGENCE

Visit us at CenterforIntelligence.org and MindMattersNews.com.

ABOUT THE AUTHOR

ROBERT J. MARKS II IS THE DIRECTOR OF THE WALTER BRADLEY Center for Natural and Artificial Intelligence, Discovery Institute; a Distinguished Professor of Electrical and Computer Engineering, Baylor University; and a Fellow of both IEEE and the Optical Society of America.

His research has been supported by defense organizations such as the Army Research Lab, the Office of Naval Research, the Naval Surface Warfare Center, and the Army Research Office. His work has also been funded by NASA, JPL, NIH, NSF, Raytheon, and Boeing. Marks has consulted for Microsoft and DARPA. He has worked in the field of artificial intelligence for over thirty years.

Marks served as editor-in-chief for the *IEEE Transactions on Neural Networks* and is co-author of the books *Neural Smithing: Supervised Learning in Feedforward Artificial Neural Networks* (MIT Press) and *Introduction to Evolutionary Informatics* (World Scientific).[1]

EXECUTIVE SUMMARY

THE DEVELOPMENT OF MILITARY ARTIFICIAL INTELLIGENCE (AI) BY adversarial countries seeking influence necessitates parallel development by countries wishing to maintain their sovereignty. History exposes as foolish today's Pollyannaish calls to ban development of lethal AI weapons. Advanced technology not only wins wars but gives pause to otherwise aggressive adversaries.

AI is often discussed without definition or foundational understanding. Despite untutored media claims to the contrary, computer programs will never be creative nor write smarter AI software. AI neither understands nor has common sense. The quest for the AI singularity of computer intelligence superiority over humanity is modern alchemy. Computer programs follow the instructions from their programs and nothing more. With these misconceptions cast aside, the true challenges of lethal AI can be accurately assessed.

Like fire, AI is neither good nor bad. The good and bad lie in human implementation. Compromising freedom, China uses AI to track the faces of political adversaries and rank its citizens. But AI can plan travel on Google maps, recognize your voice on Alexa, and save trips to the bank to deposit checks using image recognition of cell phone photos.

Most anything lethal can be weaponized using AI. The fundamental morality of the use of lethal AI lies not with AI but with the programmer and the end user. Total autonomy of lethal AI should be avoided when possible. Human supervision should be used when possible. However, in cases of communication in unfriendly environments and overwhelming attack, autonomy may be required. Human reaction time can be too slow. One large drone swarm fighting another may require a response time on the order of milliseconds.

Complex systems will invariably encounter unanticipated contingencies. A self-driving car can interpret a wind-blown plastic bag as a deer. This unexpected contingency was not anticipated by the AI programmer. As the complexity of a conjunctive AI weapon systems increases linearly, contingencies increase exponentially.

Misconceptions aside, there are still challenges with development of lethal AI. But the AI smoke is out of the technology bottle and, to maintain peace and liberty, American military AI must be developed.

INTRODUCTION

TODAY'S DOOMSDAY HEADLINES ALONE ARE ENOUGH TO TERRIFY people. "The age of killer robots is closer than you think," warns one article.[1] "We're running out of time to stop killer robot weapons," declares another.[2] Other media accounts declare that "killer robots are poised for 'MASS PRODUCTION'"[3] or scream in all caps that "'KILLER ROBOTS' WILL START SLAUGHTERING PEOPLE IF THEY'RE NOT BANNED SOON."[4]

Killer robots—also known as lethal artificial intelligence (AI) or lethal AI—utilize the powers of computer processing and algorithms to create autonomous or semiautonomous weapons that kill people. Concerns about lethal AI aren't just being raised by reporters. The United Nations has convened discussions of the threat,[5] and twenty-eight governments have already called for a ban,[6] as has the United Nations' Secretary General António Guterres. In 2018, Guterres declared that "machines that have the power and the discretion to take human lives are politically unacceptable, are morally repugnant, and should be banned by international law."[7]

More than a thousand AI researchers agree, signing a public letter warning about the dangers of lethal AI. Signatories include the late celebrated physicist Stephen Hawking, tech entrepreneur Elon Musk, and Apple co-founder Steve Wozniak.[8]

Hundreds of technology companies and thousands of individuals, meanwhile, have pledged not to participate in the development or spread of lethal AI.[9] And the group Human Rights Watch is coordinating a global "Campaign to Stop Killer Robots."[10]

The fears raised by the critics of lethal AI are grimly depicted in "Slaughterbots," a slickly produced *Black Mirror*-flavored short video, which unveils a killer drone about the size of an Oreo cookie.[11] The drone contains embedded AI in the form of facial recognition and flexible flying skills as a member of a drone swarm. The drone also contains a directed bullet-shaped exploding charge. Once programmed with the face or a characteristic of the target, the drone autonomously flies into a theater of operation and, like a honeybee fluttering from flower to flower, searches for a face match in the crowd. When found, the slaughterbot places itself close to the subject's forehead and shoots a projectile into the brain. These "slaughterbots" released in a swarm, it is argued, could win wars quickly, or be used by a rogue Republican to kill all the Democrats attending a joint session of Congress.

"Slaughterbots" condescendingly presents killer drone developers as stereotypical warmongers. Stuart Russell, a professor of computer science at the University of California at Berkeley, ends the video with an appeal to join the fight against the development of autonomous AI killing machines. The video has received more than three million views on YouTube so far.

"Slaughterbots" offers a chilling dystopian vision of the future. This vision is all the more disturbing when one realizes that the slaughterbots the video portrays are within our grasp. High tech weapons are easier to create than ever. AI tools are today readily available to those interested in making and deploying weapons. In 2015, a Connecticut teenager mounted a firearm on a small remotely controlled helicopter drone. The gun was fired remotely.[12] Drones are cheap and easy to obtain. So is software that might be used to guide them.

Paul Scharre, who was instrumental in fashioning the US Department of Defense's policy directive on autonomy in weapons during the Obama administration, agrees that "the basic concept" featured in "Slaughterbots" "is grounded in technical reality." Moreover, he warns that terrorists already have access to slaughterbot technology: "There is nothing we can do to keep [slaughterbot-like] … technology out of the hands of would-be terrorists. Just like how terrorists can and do use cars to ram crowds of civilians, the underlying technology to turn hobbyist drones into crude autonomous weapons is already too ubiquitous to stop."[13]

At the same time, Scharre dissents from the doomsday scenario put forward by the "Slaughterbots" video. "The technology shown in the video is plausible," he says, "but basically everything else is a bunch of malarkey." In particular, Scharre faults the video for promoting a number of questionable assumptions:

- Governments will mass produce lethal micro-drones to use them as weapons of mass destruction;
- There are no effective defenses against lethal micro-drones;
- Governments are incapable of keeping military-grade weapons out of the hands of terrorists;
- Terrorists are capable of launching large-scale coordinated attacks.[14]

According to Scharre, "These assumptions range from questionable, at best, to completely fanciful."[15]

Of course, the serious dangers posed by killer robots are not fanciful. Neither are the ethical challenges. But dealing with those dangers and challenges will require a sober assessment of reality, not simply appeals to emotion. This report seeks to serve as a primer for the emerging debate over the development and use of lethal AI, examining four key questions that need to be addressed if we want to create well-grounded public policies on the subject:

- Why is lethal AI necessary?
- What are the capabilities—*and limitations*—of AI?
- Is lethal AI immoral?

- What are the challenges that must be addressed if we develop lethal AI?

We'll start by looking at why the development of lethal AI is necessary in the first place.

1. The Necessity

of Lethal AI

To understand the primary reason to develop lethal AI, look no further than the history of war. History teaches that well-developed advanced technology helps win wars. New military technologies can mean the difference between life or death, between a drawn-out conflict with more casualties and more suffering and a conflict that is concluded quickly and decisively. The literature on the decisiveness of technology in war is vast. Here we will discuss just three examples. Two involve early applications of AI. The third involves the development of a technology even more horrifying than lethal AI.

Smart Bombs

Imagine a bomb dropped from an airplane guided by AI. The target is identified. The airplane's speed and outside wind speed are measured. The AI takes control of the airplane from the pilot and guides the plane to the perfect position for dropping the bomb. At just the right calculated moment the AI drops the bomb. The technology allows the bomb to drop more accurately than human control could, thus reducing unintended civilian casualties and focusing on the destruction of intended targets of military significance.

This technology sounds modern but isn't. We have just described the Norden bombsight used on American bombers in World War II (WWII). This was before digital computers but, nevertheless, the Norden bombsight used computers. The computers were not digital. They were analog. The technology for the Norden bombsight was top-secret AI and, to protect the technology, the American bombardiers were instructed to shoot holes in the equipment should a crash be imminent.[1] There is debate over whether the Norden bombsight was as accurate as claimed, but regardless, it did represent an important advance in military hardware that led to still more important innovations later.

Analog computers have fallen out of favor largely because of their inaccuracy and sensitivity to noise. Digital DVDs found favor over analog VHS tape recordings for the same reason.

Is the Norden bombsight an example of AI? At the time, yes. Today the technology is antiquated, and declassified Norden bombsights can often be purchased for a few thousand dollars on eBay.

Encryption

ONE OF Sherlock Holmes' many skills was cryptology defined as the science of coding and deciphering messages.[2] Today even the most powerful computer cannot decode modern encryption. Encryption and bitcoin secrecy shielded from scrutiny Amazon-like drug sales on the dark website *The Silk Road*. The site also sold weapons useful for terrorism, poison for the depressed and transplantable human kidneys. In 2013 the site's multi-millionaire mastermind nerd, William Ulbricht, a.k.a. the Dread Pirate Roberts, was apprehended using old fashioned detective work. Encrypted bitcoin kept financial transactions secret. Anonymous communication software Tor protected the website from outside scrutiny.[3] Neither Bitcoin nor Tor were cracked in the apprehension of the web site's mastermind.[4]

Quantum computers may soon make currently used encryption methods ineffective.[5] Effective more advanced encryption, though, will still be possible. The same quantum technology used to crack current encryption can also be used to create quantum encryption beyond the reach of quantum analysis.[6]

Encryption can be used for both for good and ill. Encryption fosters privacy necessary for liberty. But it can also be used to sell drugs or communicate messages among terrorists. For this reason, the export of encryption technology from the United States is closely monitored by regulatory agents resident at research institutions, including universities.[7]

During World War II, breaking encryption was also a priority. A version of the story is told in the movie *The Imitation Game* outlining the life of Alan Turing. The encryption decoding wasn't as sophisticated as today, but the cracking of the Nazi's Enigma machine is said to have shortened the war and saved numerous lives.

Was the mechanical machine used by Alan Turing and his team at UK's Bletchley Park an example of AI? At the time, the answer was absolutely yes. Today, the operation can be performed in a flash on a laptop computer.

The Atomic Bomb

LIKE AUTONOMOUS lethal robots, atomic bombs and nuclear weapons are chilling.

Long before the Cold War, the US and Nazi Germany were racing to develop an atomic bomb in WWII. The war in Europe ended before Germany succeeded. But suppose an American citizen-led peace movement had succeeded in banning development of the terrible bomb, and the war hadn't ended when it did? Such protests didn't happen, because the development of the bomb was kept top secret. Had the Nazis developed the atomic bomb first, flags in the US today might be sporting swastikas or big red circles on white instead of the Stars and Stripes. This is the scenario depicted in the Netflix alternative history series *The Man in the High Castle*[8] where the Allies lost WWII because the Nazis won the atomic bomb race.

The atomic bombs dropped on Imperial Japan to win WWII saved the lives of thousands of Allied soldiers, including my Uncle Junior

McHenry. Uncle Junior was a paratrooper who was trained to parachute behind enemy lines with twenty-four pounds of demolition explosives attached to each leg during the planned Japanese equivalent of D-Day.[9] Given the militancy of Imperial Japan, Uncle Junior's assignment was essentially a suicide mission. Unlike the Allied invasion of France, there were no friendly nationals behind enemy lines to help with the invasion. All of the population in Japan were trained to be hostile. The atomic bomb ended the war with Imperial Japan so Uncle Junior never had to make the jump. He returned home to West Virginia where he worked as a Greyhound bus driver, married my Aunt Justine, and raised three children. Thousands of other Allied soldiers were likewise blessed with longer lives because of the atomic bomb.

It is true that Japanese deaths from atomic bombs totalled about 146,000 in Hiroshima and 80,000 in Nagasaki, which is horrifying. But what would have happened without the bomb, which convinced Imperial Japan to immediately surrender? Historian Philip Jenkins does not paint a pretty picture of the answer:

> Invasion [of Japan] was impossible. The planned U.S. invasion of Kyushu (Operation Olympic) in late 1945 would have been one of the greatest catastrophes in military history, not least because the Japanese knew precisely where and when it was coming. They were exceedingly well prepared, with fleets of thousands of suicide bombers. The planned follow-up attack on Honshu in 1946 would never have happened because the U.S. military would effectively have been destroyed. Quite apart from the Japanese, the great Typhoon of October 1945 would have smashed the U.S. invasion fleet before it got close to the beaches.[10]

In the event of an invasion of Japan, there was a standing order to Japanese soldiers to kill thousands of American POWs. Jenkins estimates that "together with likely Japanese fatalities, you get about ten million dead—and that's a conservative figure. The vast majority of those additional deaths would have been East and South-East Asians, mainly Japanese and Chinese."[11]

The trade-off is sad, but the atomic bomb in WWII is an example of effective high technology that let more people live than die. The advanced atomic bomb technology of WWII, terrible as it was, won the war with Imperial Japan and saved lives.

As these examples show, technology is important to winning wars. The side that loses the race for technological innovation is likely to lose or at least to prolong a war, leading to more suffering, not less. What if Nazi Germany and Imperial Japan had broken our most important military codes or successfully developed the atomic bomb before we did? The outcome would have been disastrous for all—including the populations of Germany and Japan, both of which benefited from unparalleled economic and social progress after they lost the war and their totalitarian/authoritarian regimes were overthrown.

Development of Lethal AI by America's Adversaries

TODAY, POWERFUL AI technology is offered free for anyone in the world to use. As Tom Simonite reports in *Wired*, "Facebook, Amazon, and Microsoft have all, like Google, released as open source software AI that their own engineers use for machine learning. All, including to some extent famously secretive Apple, encourage their AI researchers to openly publish their latest ideas."[12]

Technical entrepreneur and maverick Peter Thiel claims that Google is "working with the Chinese military" and has been "thoroughly infiltrated" by Chinese spies.[13] China continues to steal intellectual property from the United States,[14] prompting mandated compliance officers to monitor intellectual property exports at all major US research institutions, including universities.[15] And China is developing killer robots. China's efforts in the development of AI prompt questions like "Will China lead the world in AI by 2030?"[16] China clearly recognizes the importance of technology like AI in establishing industrial and military superiority.

China is not the only country of concern. Russian President Vladimir Putin says "the nation that leads in AI 'will be the ruler of the world'"[17] and believes "that AI will offer unprecedented power—including military power—to any government that builds a big enough lead in the technology."[18] Iran ranks eighth in the world for high-impact articles on the topic of AI[19] and is reportedly building an AI supercomputer.[20] The pursuit of AI technology by adversarial Russia, China and Iran is undeniable.

No matter the degree of ambassadorial consensus, the sharing of goodwill, or the number of signed treaties, lethal AI cannot be banned. Conflicting ideologies battling for influence prohibit it. The technology is here and will be developed by those more interested in power than peace.

A nation ignores the development of new technologies at its peril. Lethal AI in America must be developed, understood, and set up as a countermeasure in order to maintain effective military defense. However much we may want to ignore or stop the technological development of lethal weapons, if we are truly concerned about our future, we have to live in the real world, and that means we need to be open to further innovation, including the development of lethal AI.

We also must reconsider our graduate-level educational policies. The United States government and state-run universities support foreign nationals from China and Iran enrolling as graduate students in high-tech programs through grants from the National Institutes of Health, the National Science Foundation, and even the *Department of Defense*. Although foreign nationals are prohibited from working on classified projects, many obtain their degrees and return to their homes with training in cutting-edge technology obtained in the United States.

Care must be taken in differentiating between potentially hostile foreign nationals and United States citizens. American citizens with a Chinese, Persian, or Russian heritage are not the concern. We learn the danger of not differentiating the difference from history. During WWII

Japanese United States citizens were incarcerated in American concentration camps. This mistake must not be repeated. Potential hostile foreign nationals require scrutiny. Not American citizens.

Many university professors are hungry for research assistants to help in the attraction of funding and the writing of journal papers. Fully funded researchers are especially welcome. There were more than 360,000 undergraduate and graduate foreign national Chinese students in the United States in the 2017-2018 school year.[21]

I currently work on unclassified projects on developing the next generation of cognitive (smart) radar, funded by the Department of Defense. Participation of foreign nationals in this research is allowed. Our publications about the research in the open literature attracts foreign nationals, and I periodically receive emails from China and Iran asking to study with me in my research field. Most are fishing for any university support that would get them into a United States university. Some hopefuls are fully supported with scholarships from home. I recently received the following email from China:

> The research topic of my doctoral [project] is cognitive radar waveform design …. I am really interested and hope that I could have the opportunity to study and work under your supervision. My living and plane ticket costs will be covered by CSC Postgraduate Study Abroad Program.

Likewise, an email to me from Iran reads:

> My … [graduate engineering] thesis was on [the topic of radar]. In my thesis, I have some articles about my research. In addition to, I'm Coauthoring in two books [on the topic]… I would be highly honored to become a member of your research group and to make a common research work under your supervision.

The development of smart radar is part of a larger effort in application of AI in the military. The goal is to develop powerful technology with precise lethality and to give adversaries pause when considering conflict. Thus, even students in non-classified areas of research can gain expertise in fields directly related to lethal AI.

But what exactly *are* the capabilities of lethal AI? And what are its limits? In the next section, we will examine what we mean by artificial intelligence, summarize its usefulness in weaponry, and describe some of the limits to AI that are often not covered in public discussions.

2. AI's Capabilities

and Limits

THE TERM ARTIFICIAL INTELLIGENCE OR AI IS OFTEN USED WITH no reference to a definition. Eliezer Yudkowsky states: "By far, the greatest danger of Artificial Intelligence is that people conclude too early that they understand it."[1] What is AI and its capabilities?

In scholarly venues there is teasing apart the disciplines of artificial intelligence, machine intelligence, and computational intelligence. In the media the term AI refers to any astonishing task involving the use of computers. Operation is often enhanced by wireless communication and other technologies. This definition of AI thus includes Alexa, Uber, Lyft, Google maps, search engines, natural language processing, Netflix, Amazon shopping, bitcoin, Dragon voice recognition, Wikipedia, Skype, and Dropbox. These technologies are proven. They do not live in the world of speculation or journal papers. As time passes, familiarity numbs awe. Email and even calculators were once considered cutting edge AI.

AI, like artificial neural networks, is not new. Artificial neural networks are a type of AI so named because of a crude relationship of the software architecture to the wiring of the human brain. The human brain has about 100 billion neurons[2] that communicate with each other through synapse interconnections. Although the number of computer simulated neurons is much smaller, artificial neural networks likewise have connected neuron-like nodes. Training a neural network involves tweaking the strength of the connections with the goal of morphing the network into representing the database used for training. These type of neural networks are called layered perceptrons.[3]

Deep convolutional neural networks are a special case of the layered perceptron. Layered perceptrons are classically trained with *features*. Consider training of a neural network to differentiate a sumo wrestler from a basketball player. The obvious features that differentiate sumo wrestlers from basketball players are height and weight. These two features are gathered from a number of subjects and used to train a simple neural network. Then, given the height and weight of a subject not used in the training, the layered perceptron will announce whether the candidate is a sumo wrestler or a basketball player.

Deep convolutional neural networks, however, can be trained directly from the pixels of an image. Basketball players and sumo wrestlers need not be characterized by features. The pixels from pictures of sumo wrestlers and basketball players can be used.

An example of what happens in a convolutional neural network is illustrated by the physiology of the dragonfly. The dragonfly has 24,000 ommatidia light receptors in its eye, each providing a single picture element.[4] To detect whether it is flying over water, the dragonfly's neural system begins layer by layer to discard more and more unwanted information until a binary decision is reached. Is the light incident on its optical receptors polarized or not? If polarized, the dragonfly knows it is flying over water. Otherwise, it is over land.

This is descriptively what happens in a deep convolutional neural network. Layer by layer, unnecessary information is discarded until a binary decision is made. Is the input candidate a sumo wrestler or a basketball player? The process is called *deep* because a large number of processing layers are used. The *convolution* term refers to a mathematical operation that occurs at each layer of the deep neural network.[5]

Reinforcement learning is a key component in the training of AI to win the board game GO. Reinforcement learning does not use data but, in the learning phase, explores the many paths of a problem solution again and again until a highly successful solution is found. For a given setting of stones on a GO board, there are many moves. Which is the best?

All AI requires the human tweaking of parameters. Reinforcement learning needs to be tuned between the tasks of improving old solutions and considering new paths. Layered perceptron neural networks require tuning of parameters such as learning rate. All successful AI needs human guidance to perform successfully.

Both deep convolutional neural networks and reinforcement learning are subsumed in the general category of *deep learning*. Deep learning is made possible by powerful computers that, while learning, can tirelessly crunch numbers in a manner previously not possible.

Historical Successes

ARTIFICIAL NEURAL networks have chalked up a number of successes since their introduction in the mid twentieth century. In 1960, Stanford Electrical Engineering professor Bernard Widrow used neural networks to forecast weather, win at blackjack, translate speech to the written word, and balance a broomstick.[6] The control on a Segway personal transporter[7] performs an operation equivalent to a broom balancer. Widrow's neural network's weather forecast was better than a local meteorologist and the success of blackjack was close to its theoretical limit. Widrow's neural network was used in a control application in the supersonic commercial jet airline Concord. About the same time, Bell Lab's Claude Shannon was teaching a robot mouse how to run through mazes.

Successful commercial applications of neural networks emerged in the last decade of the twentieth century. Neural networks were widely used to forecast load demand in the power industry.[8] Robert Hecht

-Nielson used neural network-based AI to check credit fraud and sold his business to Fair-Isaac of FICO fame for over 800 million dollars in 2006.[9]

Deep learning is the secret sauce behind the astounding victory of AI over the world champion in the board game of GO. Historically AI was able to sequentially master the games of tic-tac-toe[10], checkers,[11] chess,[12] and then GO. Each game was more difficult than the next. Deep learning was then applied to winning Atari arcade video games only using pixel information from the video display of the game.

From missiles to drones, AI can be used to augment the operation of most any weapon to the status of a killer robot. (The human shapes normally associated with robots often have nothing to do with the underlying AI.) The question in these amped-up weapons is less about their degree of lethality and more about the degree of allowable autonomy. Killer robots under the control of a human are a simple augmentation of warfare capability. But there are situations where complete killer robot autonomy is necessitated, which has caused some to worry that the power of AI will become unlimited.

The Limits of Lethal AI

AI HYPE has led to fear that AI will create a race of super-human robots that will ultimately replace us, or at least be completely uncontrollable by us. Despite AI prophecy often bordering on the hyperbolic, the following limitations of AI are becoming more evident:

- AI cannot be innovative or creative. Among those who agree are Oxford's Roger Penrose[13] and Johns Hopkin's Gregory Chirikjian.[14] The inability of AI to be creative is specifically addressed by Microsoft CEO Satya Nadella[15] and Yale's David Gelernter.[16] The AI singularity theory purports that AI will someday write better AI that writes better AI to the point where AI exceeds the capabilities of humans. This fanciful prophecy incorrectly assumes AI can be creative.

- AI lacks common sense and requires continual sanity checks. When at all possible, trained AI results must be checked by a human with appropriate domain expertise. Any AI medical diagnosis should be checked by a doctor if possible. Recommendations made by AI in any military theatre should be approved by command. Lethal weapons based on any type of AI require human oversight when at all possible.

One of the most famous cases of AI in popular culture is IBM's Watson, originally created to answer questions on the television show *Jeopardy*. To great acclaim, Watson eventually won a competition on *Jeopardy*. Yet its programmers were concerned about Watson's lack of common sense prior to *Jeopardy*[17]: "The IBM (Watson) team was afraid the *Jeopardy* staff would write clues with puns and double meanings that could trick Watson." Simple phrases like "the missile cannot destroy the bridge because it is too big" can be incorrectly interpreted by AI to mean the missile is too big. To see this, read the sentence again like a computer: "the missile cannot destroy the bridge because it is too big." Because of its lack of common sense, an AI expert has called IBM's promotion of aspects of Watson "a fraud." A business investor called Watson "a joke."[18]

Let's drill down. Is deep learning applied to winning board games extendable to real world scenarios like mastering war games or managing economies? Can the data -mining capability displayed by Watson be applied to rummaging through the latest medical literature to help physicians provide better treatment for cancer patients? Attempts to apply much cutting-edge AI outside of their silos of success have not been successful. A major contributing factor is AI's lack of discernment, creativity, and basic common sense.[19] As Roger Penrose put it, "Intelligence cannot be present without understanding. No computer has any awareness of what it does."[20]

Learning and Ergodicity

DEEP LEARNING has also suffered setbacks in applications outside of board games. One reason is AI's black box affliction. The reason for a neural network's decision is difficult if not impossible to connect to a cause. A naïve neural network trained only to differentiate between friendly tanks and enemy tanks will classify a kumquat as either a friendly or enemy tank.[21]

AI might learn in ways other than anticipated by the programmer. A neural network trained to differentiate between dogs and wolves was found, in the case of wolves, to classify according to the presence or absence of snow in the background of the wolf picture rather than by the features of the wolf.[22] Deep convolutional neural networks in particular suffer from undesirable sensitivity. In some cases, deep convolutional neural networks can be fooled by changing only one pixel in an image.[23]

AI is restricted to AI ergodic problems.[24] In mathematics, ergodicity comes in many flavors. AI ergodicity is the simple property that experiences from the past can be used to forecast events not yet seen. Future values of power load can be forecast using values in the past. A deep convolutional neural network trained to differentiate between a basketball player and a sumo wrestler from picture pixels can be used to classify subjects not yet seen.

AI ergodicity requires time invariance. A scenario evaluated in the past and used to train must be the same as that considered in the future. But time invariance is not sufficient to establish AI ergodicity. Flipping a coin gives the same probability every flip and is time invariant. But the history of coin flips will not be helpful in forecasting the next coin flip. Coin flipping is not AI ergodic.

Casinos began posting the immediate past winnings of the spin of a roulette wheel in hopes of fooling dumb players. Black has won five times in a row in the five previous games, they reason, so red is overdue on the next spin. Players, thinking roulette wheel outcomes are AI ergodic, bet on red. The spinning of the roulette wheel is time invariant. The same game is played again and again without variation of the rules. But the future cannot be forecast from the past. Since the future cannot be determined from the past, roulette is not AI ergodic.

A classic example of a process that is not AI ergodic is stock market data. The tick data from the past cannot be used to viably forecast stock market values in the future to make money.

Extreme time-variant phenomena cannot be captured by AI.[25] AI trained to classify caterpillars will be of little use for later tracking the butterflies they become. Classifying the metamorphizing caterpillar from caterpillar data alone is obviously not AI ergodic.

A major hindrance to application of AI in adversarial conflict is disruption of an intelligence's ergodic assumption. An adversary will attempt to disrupt any ergodicity assumption, thereby destroying the accuracy of the AI.

Doing so need not be high tech. AI facial recognition can be disrupted by shining pocket lasers into the camera lens. This tactic was recently used by Hong Kong protesters to protect their identity from the Chinese government.[26] More subtly, jewelry can be used to fool facial recognition software.[27] AI radar trained to detect aerial decoy flak may not work well if the flak type is significantly changed. Such masking is not new. In WWII Allied forces used inflatable or wooden dummy tanks to fool Nazi aerial reconnaissance. Today, mobile tanks can alter their appearance to fool enemy AI aerial reconnaissance trained narrowly to only detect undisguised tank images.

Non-adversarial AI applications do not prompt gaming of the AI. A deep neural network trained in image recognition assisted in the locating of a crashed helicopter.[28] AI trained on multiple images of helicopters is addressing an AI ergodic problem. During application after training, the AI ergodic assumption was not violated.

Google-owned DeepMind is the company behind AI mastering the board game GO. Board games like GO are strongly AI ergodic. The layout of stones on the board changes, but the same game is played using the same rules again and again (and again). Like Watson, DeepMind has yet to show notable success in otherwise applying the technology used to train board and arcade games. In 2018, DeepMind lost $570 million.[29] The company claim is that more effort was placed into recruitment than on development. In looking for a major commercial application breakthrough, Google CEO Sundar Pichai spins: "Looking at the pace of progress, I think we will have AI in a form in which it benefits a lot of users in the coming years, but I still think it's early days, and there's a long-term investment for us."

Even if there are sharp limits to what AI can do, it is still powerful, and it has the potential to dramatically change warfare as we know it. The US military is infusing proven AI into electronic warfare,[30] including image intelligence in aerial surveillance and application of machine intelligence to develop cognitive (smart) radar.[31] Such development contributes to lethal capability in conflict.

Some claim that the impact of lethal AI will be so dramatic and terrible that the technology is inherently immoral and therefore its development must be stopped at all costs. In the next section, we will explore whether they are right.

3. The Morality
of Lethal AI

THERE IS MUCH TALK TODAY ABOUT THE SUPPOSED IMMORALITY OF the military use of AI. As a result, employees at technology companies are increasingly protesting their companies' involvement with military contracts related to AI, even those that are not directly involved with the creation or production of "killer robots."

Project Maven is an AI project funded by the Pentagon to help the US military develop computer applications that can autonomously identify "objects of interest from moving or still imagery."[1] More than 3,000 Google employees protested the company's involvement in Project Maven, and under pressure, Google dropped its involvement in the project.[2]

Similarly, Microsoft was awarded a $480 million contract to provide their HoloLens to the US Army.[3] The HoloLens allows soldiers to view information in visors without the distraction of looking down or away. The technology is called *augmented reality*, a term coined in 1990 by Boeing engineer Tom Caudell.[4] Microsoft workers protested, saying "We are a global coalition of Microsoft workers, and we refuse to create technology for warfare."[5]

This knee-jerk reaction against the use of AI applications for the military is misguided. Discussing all relevant ethical questions relating to military use of AI lies beyond the scope of this report, but here are a few key points.

AI Applications Have a Range of Military Uses

MANY OF those protesting the development of AI applications for the military seem to think that the only real military use of AI is to create autonomous killer robots to kill more people without human direction or oversight. That is simply not the case. In fact, one of the most important military roles for AI is to help humans make better decisions that can help save innocent lives. For example, soldiers entering a densely packed urban area to clear out terrorists can use AI-based systems to better distinguish between real and fake threats, protecting soldiers' lives as a result and reducing the harm to innocent civilians who may otherwise be harmed by mistake. The HoloLens can be used in field surgery to save the lives of those wounded in combat. It is hard to defend the view that use of AI to save lives and reduce civilian casualties is immoral.

Lethal AI Is Morally Defensible

MANY PEOPLE were killed by the Allies in the fight to liberate Europe and Africa from Nazi Germany. Most people think the killings were morally justified. That is because most people are not pacifists, and most people understand that at least some killings can be justified to reduce more horrendous killing or defend one's fellow citizens or neighbours from harm. WWII was a just war.

In warfare scenarios, AI can be more morally defensible than other military technologies due to the fine-tuning of targeting. Just war can require military action to avoid, as much as possible, harm to non-combatants and their property. The horrific bombing of Dresden in WWII resulted in a 1600 acre firestorm, killing over 20,000. Technology today can be used to control long-range precision-fire missiles with pinpoint accuracy, thus reducing non-combatant deaths.

Morality of Lethal AI Belongs to Its Designer and User

IN THE end, AI has no more resident ethics than a toaster. A toaster plugged into the wall can be used to make toast or can be thrown into an occupied bath to electrocute the occupant. Technology is universally neither good nor bad. It's how it's used. AI used to counter evil AI in just wars is both good and ethical.[6]

Ethics does not belong to AI but rather to the programmer and user. A classic ethics dilemma is the trolley problem.[7] A runaway trolley is moving toward a small car with five people locked inside. Three of them are babies. A collision will kill them all. There is a lever that controls a rail switch. If the lever is pulled, the trolley will be redirected and the five people in the car will be saved. But there is a problem. If the switch is flipped, the world's leading cancer researcher, who is tied to the other track, will be killed. Should you pull the lever or not?

Both sides of the trolley problem can be debated *ad nauseum*. But one issue is clear. If the lever is controlled by autonomous AI and the AI makes the decision, responsibility for the outcome does not lie with the AI. It lies with the computer programmer who wrote the AI code. Asking whether AI has ethics is like asking a toaster whether it likes white or wheat bread.

The expertise and goals of the programmer must be translated into the machine language of AI. If the programmer is evil, the AI will perform evil tasks. If the programmer is stupid, their AI can do stupid things. And no matter how good the programmer is, there will be unexpected results. Not all contingencies can be imagined. If the unexpected result is not because of carelessness or stupidity, it is called an accident. And accidents happen.

The Moral Use of Fear

"To BE prepared for war is one of the most effectual means of preserving peace."[8] This classic saying from George Washington points to a truth we would all do well to remember. The highest purpose of weaponry like lethal AI is not to carry out a war. It is to prevent war. One of the sad realities of the human condition is that fear of someone else's weaponry can make war less likely, not more. So developing a powerful weapon may be more likely than disarmament to achieve the goal of peace. As

Theodore Roosevelt famously said, "Speak softly and carry a big stick." And, we would add, let adversaries know about the big stick.

Here is a twentieth century illustration. Ronald Reagan's Strategic Defense Initiative was meant to develop defensive technology to prevent a nuclear attack. While I was at the University of Washington in 1988, I received a grant from the Office of Naval Research related to Ronald Reagan's Strategic Defense Initiative (SDI), better known as "Star Wars."[9] SDI's aim was to shoot down threatening missiles aimed at the United States using space-deployed technology. Many of my naïve idealist ivory tower colleagues at the University of Washington refused to participate in SDI. They claimed the underlying idea was too evil. They were also not happy I was involved. The SDI had to do with war, they reasoned, and all war was bad.

The SDI program, though, seriously scared the Soviets. The fear leveraged the General Secretary of the Communist Party Mikhail Gorbachev to meet with Reagan at Reykjavík, Iceland in October 1986.[10] Gorbachev wanted the US to end SDI so badly, he offered Reagan sacrificial concessions in their negotiations. Reagan refused to call an end to the SDI program and no deal was reached.

Subsequently, SDI turned out to be instrumental in the collapse of the Soviet Union and the ending of the cold war—without a shot being fired. The Soviet Union figured it would go bankrupt trying to keep up with the US defense budget. The good intentions of my anti-war-at-all-costs colleagues, if adopted, would have prolonged the Cold War.

Anyone who understands the moral use of fear to prevent war should also understand why weaponry such as lethal AI can be defended as a strategy for preventing war.

In sum, there are plenty of moral justifications for continuing to develop lethal AI. The main challenges to adopting lethal AI are not ethical. They are ethically practical. In the next section, we will discuss some of them.

4. THE CHALLENGES OF

IMPLEMENTING LETHAL AI

ONCE IT IS AGREED THAT LETHAL AI SHOULD BE PURSUED, THE policy debates don't end. In fact, they will be just beginning. The hardest decision isn't whether to develop lethal AI, but how to handle the practical issues that will arise once development is seriously underway. In this section, we will look at four key challenges that need to be faced if we develop lethal AI, along with some strategies for dealing with them.

Deciding How Much Autonomy to Provide AI Weapons

PROBABLY THE most controversial question we face in developing lethal AI weaponry is how much autonomy to provide. Although media coverage of "killer robots" often treats all AI weapons together as uncontrolled by humans once they are unleashed, this is inaccurate.

Semiautonomous AI weaponry has humans in the loop. Hence the prefix "semi." This includes base station control of outfitted missiles with onboard cameras, and the launching of loitering munitions from submerged submarine platforms. There is less controversy about semiautonomous weapons because human judgment is always in control. Humans should be involved in the assessment of AI decisions when at all possible.

Contrast this with totally *autonomous AI* weapons where no human is involved. Once deployed, autonomous weapons make decisions on their own, independent of human counsel. If an observer has a finger posed over a self-destruct button during operation, is the autonomous machinery now semiautonomous? It depends on your dictionary. The matter is even more confused when the terms semiautonomous and autonomous are redefined for publicity and political reasons.[1]

There are degrees of autonomy. A drone swarm, for example, might have a number of totally autonomous operating modes. A human can switch an otherwise autonomous swarm mode from defensive to offensive. A human is in control, but at a high level. The swarm mode might also be controlled by a human turning a knob rather than flicking a switch. This happens in social insect swarms but in an autonomous fashion. Worker ants will switch their roles from worker to army ants when their anthill is being attacked. The greater the perceived danger, the greater the number of conversions. The adaptation is gradual. For ants, the transition occurs without external control. The threshold of conversion activation is spread throughout the swarm and different ants transition roles when their individual threat threshold is exceeded. For automatous military drone swarms, the transition knob can be turned by a human. Indeed, a number of tunable parameters for semiautonomous AI can each be individually controlled by knobs.[2]

The *Aegis Combat System* is an example of a weapon with lots of knobs and switches. Aegis, a smart and powerful naval weapons system, tracks and guides weapons to destroy enemy targets. The system is described as "a dangerous dog kept on a tight leash."[3] The operation of the Aegis can be operated at different levels as determined by human oversight.[4]

The important distinction in the fuzzy region of semiautonomous and autonomous operation is the degree to which a human has control of the AI. Although ultimate human control is a worthy goal of developers of AI weapons, there are times when this is not possible. Required reaction time might exceed the abilities of humans.

Two battling drone swarms can have numerous agents who, in order to be effective in combat, individually require reaction times in the milliseconds. Humans cannot react quickly enough for one, let alone hundreds, of interacting swarm agents. Autonomous operation can be appropriate.

Being overwhelmed can necessitate autonomy. Anyone who has played the 1978 Arcade game *Space Invaders* can relate. In the beginning of the game, rows of attackers move slowly and predictably back and forth across the top of the screen. If not destroyed, the attackers also move incrementally closer to you, the shooter, until they are on top of you. When the game is slow in the beginning, the shooter can aim and shoot the invaders individually. Once the first wave of invaders is destroyed, a second faster group starts bombing while moving more quickly. Ultimately, the attackers become so fast there is no longer time to aim. The best one can do is spray the many attackers with a barrage of bullets. If speed continues to increase, no matter how good a player, there will come a point where human reaction time isn't fast enough. Total autonomy is one answer.

Consider then, being attacked by a large hoard of missiles all traveling at supersonic speed. There is no time to respond in a careful, methodical manner to each missile. An autonomous action can be the only viable response option. Computers, not constrained by slow human reaction times, can assign antimissiles to each attacking missile and win the day.

Such military weapons exist: "More than thirty nations already have defensive supervised autonomous weapons for situations in which the speed of engagements is too fast for humans to respond."[5]

Autonomy is also required in other cases.

Control communication with deployed AI can be interrupted, thereby leaving the AI on its own. Indeed, disruption will be an objective of the enemy, who will try to jam communications with misleading signals, rendering friendly control impossible. The experimental X45 uninhabited autonomous aircraft developed by Boeing was designed with this in mind.[6]

Even the use of signals to communicate with unmanned AI aircraft can be dangerous. Control signals can be detected and localized by the enemy and used to pinpoint and destroy the control center. Homing in on the source of radar signals by the enemy can result in the missile destruction of the radar facility. The Israeli-developed *Harpy* is a missile designed to do just that.[7]

There are other scenarios where autonomous AI is necessitated when a control signal is lost.

A drone may be sent into an enemy building to fly about and provide a map of the inside. Two or more drones "can explore, collaborate, and gather intelligence in their environment" inside the building.[8] If there is no communication, structure information must be stored in the drones and retrieved later.

An armed robot or drone exploring winding cave-like structures for enemy combatants may be deprived of communication by its environment. Like walls diminish the WiFi signal in your home, radio waves are weakened when they go through walls. Thick, damp, rock cave walls protected by wet soil can likewise attenuate radio strength enough to make communication impossible.

Autonomy is often necessitated in deep water. Underwater vehicles like submarines are limited to acoustic (sound wave) communication, which is extremely slow. Radio waves travel in water about as well as a laser pointer's beam goes through chocolate milk. Underwater environments can offer a large degree of autonomy. Submarines can be nearly undetectable when submerged and are therefore difficult to locate and destroy. Autonomy for unmanned autonomous underwater vehicles (AUVs) is therefore often required.[9]

AUVs have many non-military uses. They are used for oil exploration, surveillance, underwater pipeline inspection, and environmental monitoring. The AUV is also a great way to smuggle drugs across waterways, if you can afford it. The military uses AUVs for defensive purposes such as surveillance and mine detection.

Armed AUVs can be used to provide a chilling lethal punch. Consider a fleet of almost undetectable nuclear-armed AUVs loitering in deep water. They keep slowly moving to escape detection. When a short acoustic code is heard, the AUVs surface and launch their lethal payload. How can such an enemy weapon be countered? Counter patrolling AUVs tasked with searching for hostile AUVs will help. A more effective answer is not available but will undoubtedly involve development of new technology.

It is always advisable to have a human in the loop for lethal AI operation. However, there are cases when this is not possible. If quick action is needed in complex scenarios, total autonomy might be the only viable option. Whether or not to apply autonomy will itself remain the decision of humans.

Planning for Unintended Consequences

DEVELOPMENT OF AI to be used in life-threatening situations must be thorough. The major problem with AI is that even the best computer programmers can't think of everything. There will always be contingencies outside of their consideration. AI is brittle. AI will do things the programmer has thought of and things she didn't. AI responding to unplanned situations can have catastrophic consequences, so planning for the unplanned is another key challenge for the development of lethal AI.

Consider the following example. Nazi U-Boats in WWII used acoustic sensing torpedoes. The torpedoes listened for engine noise and zeroed in on the target. The launched torpedo would detect and aim according to loud noises made by Allied ships as detected by the torpedo. At least that was the plan. The problem was that the U-Boat itself had an acoustic signature. The U-Boat made noise. Once launched, an acoustic sensing torpedo might hear noise behind itself, turn around, and blow up the U-Boat that launched it. To avoid suicide, the U-Boats began to turn off their noisy engines after a torpedo launch. Later, the torpedo technology was improved. The torpedo's acoustic sensor was changed to not activate until the torpedo was far enough from the U-Boat. Far enough away, the U-Boat engines would be an undetectable whisper. In such ways a Band-Aid fix can be added to individual unexpected technical results. But other glitches can be right around the corner.

Computer code does what you tell it to do—not necessarily what you want it to do. Ideally, they're the same thing. Anyone who has written and debugged software has experienced software brittleness. You write some code, then run the program to see if the code is doing what you want it to do. Something unexpected happens. You look at the code and say "Of course! I wrote such-and-such a line and the program did exactly what I told it to do. I didn't mean that. Silly me." What you told the program to do was not what you wanted it to do. So you go back and change the code so that things work more like how you want them to.

No matter how well meaning, morality imposed on AI can itself have unintended consequences. Consider the following lofty moral AI law from science fiction writer Isaac Asimov: "A robot may not injure a human being or, through inaction, allow a human being to come to harm."[10] It sounds great. But all possibilities must be considered before adopting. Here's an example. A mass shooter enters a church and begins shooting parishioners using an AR-15 with a bump stock and multiple high-capacity magazines. The gunman, dastardly deed completed, exits the church. A police officer is outside and unholsters a stun gun. The officer's goal is to incapacitate, secure, and then arrest the shooter. A nearby robot observes the action and remembers the command: "A robot may not injure a human being or, through inaction, allow a human being to come to harm." The shooter is a human. So the robot deflects the police officer's stun gun and the shot goes wild. While the officer unsnaps another holster to remove her Glock, the killer gets away and is never captured.

The well-meaning robot in the church shooting incident is an example of brittleness where the original no-harm moral guideline needs to be amended. The amending itself cannot come from AI, but must come from the human programmer.

Such detailed guidelines are the stuff of lawmaking where, ideally, all possibilities and special cases are considered. Considering all cases, though, is never possible to a certainty. The policymakers can only try and do their best. That's why we have courts of law who dicker about the fuzzy areas of legal policies on which the law is not crystal clear. AI will never be capable of making such judgmental distinctions outside of what it is programmed to do.

Windblown plastic bags are the urban tumbleweed. A self-driving car mistakes a windblown plastic bag for a deer and swerves to miss it.[11] After making this mistake, the AI can be adapted so as to not repeat the mistake. The problem, of course, is that all contingencies cannot be anticipated. There will always be some other unexpected occurrence. As a result, totally autonomous self-driving cars will always be put into situations where they will kill people. Should such cars be banned? The answer depends on how many people they kill. Human-driven vehicles have never been outlawed even though human drivers also kill. So totally autonomous self-driving cars might be adopted for mainstream use when they kill significantly fewer people on average than human driven vehicles.

Unintended consequences of complex autonomous AI will always be present. The role of the programmer is to minimize them.

Ensuring Adequate Testing

A MAJOR problem with total AI autonomy is testability. Will the AI perform well in all possible contingencies? To answer the question, the AI must be examined and tested for both program glitches and unanticipated situations.

Consider the land mine. The typical land mine is buried in the dirt. It explodes when a simple pressure sensor is activated. Notice that the simple land mind fits the definition of an autonomous weapon. There is no human in the loop to decide whether or not the land mine explodes. Once the land mine is deployed, it is totally on its own. The effects of a land mine are devastating, but the action of the device is simple and well known to those who plant them. Little testing is needed to determine what the land mine will and will not do. When deployed, most all contingencies of its operation are understood.

The land mine illustrates that the simpler an autonomous system, the easier it is to effectively test. The land mine is not complex.

There are many proposed procedures for measuring complexity.[12] All conclude that high complexity increases contingencies, thereby increasing security vulnerability and the existence of unanticipated consequences. This is intuitively obvious. Less obvious is that a linear increase in complexity can cause an exponential increase in contingencies.[13] The relationship is much worse than mere proportionality. The exponential increase makes testing more and more difficult as complexity increases. This suggests that AI weapons with narrow missions can be more easily scrutinized before use. In the case of the ability to perform multiple missions, combination of subtasks should be disjunctive rather than conjunctive.[14] This is true of the overall highly complex *Aegis Combat System* where human operators choose among a plethora of available narrow actions.

Bullets, arrows, and even thrown rocks are simple fire-and-forget weapons that, once launched, operate autonomously. Fire-and-forget missiles, like the Boeing's Harpoon[15] anti-ship and land missile,[16] uses GPS, radar, and image recognition to identify and destroy a specified target. The mission of a launched Harpoon missile is simple and narrow.

Historical mistakes made by Harpoon missiles are chalked up to human error.

Testing of autonomous self-driving cars is possible because of the wide variety of available test scenarios. Using humans with interrupt abilities, miles of test driving under all sorts of conditions is possible. The test is performed in an AI ergodic established environment in a peaceful theater of operation. Once a car is trained to drive over interstate highways, we can expect that car will be useful over interstate highways in the future. The highways don't change much. They are AI ergodic. Except for possible terrorists, no one is trying to game the system and make cars crash.

Similar testing is needed for autonomous military weapons. But the situation is more difficult. There are plenty of roads on which to test and tune the self-driving cars, but there are not a lot of wars available to test and tune autonomous AI weapons. War games, field tests, and simulations must suffice.

Constructing contingencies to test is the job of military tacticians. Imaginative and creative brains are needed to assess all possibilities. This requires bright minds able to conceive of nearly all probable contingencies. For complex systems, anticipating all contingencies will never be possible. And the more complex the system, the more difficult the establishment of reliability and the greater the number of unanticipated vulnerabilities. This is inescapable in the testing of highly complex autonomy. As self-driving cars will always kill people, autonomous weaponry will always have a chance of making a mistake. Meta-analysis by humans mitigates this. It would be nice if there was a super AI with the ability to analyze AI shortcomings to the degree needed. But computer programs famously lack the ability for sweeping analysis of other computer programs.[17] Testing is and will always remain the responsibility of humans, either directly or using expert system computer code.

The need to analyze a wide spectrum of contingencies becomes evident when considering military tactics. An intelligent enemy will attempt to anticipate the abilities of your AI. The enemy wants to make your AI ineffective. As has been the case throughout history, the enemy will try to anticipate a contingency not yet considered and use it to make your technology ineffective. Such vulnerability can increase exponentially with complexity.[18]

Once self-driving car software is tested and established, the design is basically done and no one is trying to kill you. Not so for military applications. When measures to disrupt your AI are discovered by the enemy, you must begin development of countermeasures to make the AI effective again. The back and forth will continue if unchecked by policy or treaty.[19]

Developing Countermeasures

THOSE OPPOSED to lethal AI worry that once developed, deployment will be unstoppable and we will be completely at AI's mercy. As frightening as this doomsday scenario is, the history of technological innovation in both business and warfare suggests another more likely possibility.

History is replete with accounts of new military technologies trumping old. First, there were military airplanes. Then there was radar as a tool to shoot them down. Then there was stealth technology to avoid radar. Evil, seeking influence, demands a response, so the technology to provide a response must be developed. During the Cold War, technology acceleration was called the *Arms Race*. First, there were missiles. Then there were anti-missiles. Then there were anti-missile-missiles. In our own day, Israel has deployed a sophisticated missile protection system known as *Iron Dome*.[20]

Instead of acting like the proverbial Chicken Little when it comes to lethal AI, effective countermeasures require development. The notion of helplessness against killer robots, military expert Paul Scharre insists, is a "farce": "Every military technology has a countermeasure, and countermeasures against small drones aren't even hypothetical. The U.S.

government is actively working on ways to shoot down, jam, fry, hack, ensnare, or otherwise defeat small drones."[21]

Let's just look at one particular case where countermeasures could be employed to counter the threat of lethal AI. Without a doubt, swarming clouds of autonomous killer drones promise military effectiveness. Swarms are robust.[22] Kick over the dirt mound of a fire ant hill and stomp the swarming ants. No matter how well you stomp, the swarm survives and lives to build another anthill. And if you stomp too long, a surviving ant will eventually bite you on the ankle. Likewise, half of an armed drone slaughterbot swarm can be destroyed and those surviving can still be a threat.

Swarm tactics in the military are not new and date as far back as Alexander the Great. The swarm agents then were mounted archers on horseback: "[The] Persians were able to conquer a vast Middle Eastern empire, one that Alexander the Great conquered only after developing his cavalry's own counterswarming capabilities. The Persians recovered their empire soon after Alexander's death, continuing their military tradition of swarming—later using it to destroy the Roman legions."[23]

Is there an effective countermeasure to swarming? Alexander's swarm countermeasure forced the swarm towards a barrier like a mountainside or the edge of the ocean. Once cornered, the swarm has no room to maneuver and is more easily destroyed. Alexander's swarm countermeasure remains effective if applied to drone swarms today. A swarm of slaughterbots can be contained by something as simple as chicken wire.[24]

All sides will be seeking to develop countermeasures to their opponents' lethal AI weaponry. But that is why AI and other technologies must continue to be developed with never-ending vigilance in order to counter current and potential military and terrorist threats.

5. Living in a World of Lethal AI

Can we live in a world of lethal AI?

We don't have a choice. Like it or not, AI is here to stay and will be developed by the good and the bad.

Autonomous weapons, already in the America's arsenal, give enemies pause. Defensive AI weaponry is needed to counter enemy AI weapons which will undoubtedly be built. Treaties and agreements, if reached, help but do not solve the challenge.

AI is changing quickly and must continuously be monitored. Doing so requires separating the wheat of usefulness from the chaff of hyperbole and the uninformed. As the wave of AI continues to surge forward, proven and useful technology is left in its wake. The final proof of the value of any technology is its reduction to practice in industry, the military, or commercial products. Vetting continues on popularized AI tools such as Watson-like data-mining oracles and deep learning, including convolutional neural networks and reinforcement learning. On the other hand, proven AI computer technology from face recognition to the tuning of voice recognition software is now ubiquitous.

A technological arms race with regard to lethal AI might develop to a standoff, at least for a time. Thus, one consequence of the development of lethal AI might be the frightening—but effective—strategy of mutually assured destruction (MAD) normally associated with thermonuclear weapons. The use of hydrogen bombs is seen as so horrible today they are no longer even tested. Similarly, the use of chemical and biological weapons on the battlefield is banned by treaty and has not been used by the world's major powers against each other since World War I.

But even in this case, work on countermeasures needs to continue. Despite agreements among the world's major powers, there are always outliers, which is why brash rogues like North Korea's Kim Jong Un will try to build atomic bombs and threaten humanity with them. And murderers like Iraq's late Saddam Hussein and Syria's Bashar al-Assad will continue to kill with chemical weapons.[1] Treaties and agreements are only of use among the honorable. Ask Neville Chamberlain about his "peace for our time" treaty with Adolf Hitler.

So how do we live in a world where AI is tasked to kill? In the same way humans lived in previous generations when confronted by fearful new military technologies, whether it be the cannon, the Gatling gun, chemical weapons, or the atom bomb. Alfred Nobel funded the Nobel Peace Prize in part because of concern that his invention of dynamite would accelerate the world's destruction. All such threats still exist but have been largely contained.

America needs to continue to harness the powers of human ingenuity, adopt a realistic assessment of human nature, and be guided by the time-honored codes of morality. A sober assessment of the motives and capabilities of adversaries is mandatory for long-term survival. Technological development is a mark of our humanity, and we have faced this situation before. In the face of future technologies not yet known, we will undoubtedly have to face it again.

ENDNOTES

ABOUT BRADLEY CENTER
1. The Walter Bradley Center for Natural & Artificial Intelligence, https://centerforintelligence.org/about/mission/.

ABOUT THE AUTHOR
1. This document does not necessarily represent the views of and has not been reviewed or approved by Baylor University.

INTRODUCTION
1. Kelsey Piper, "Death by Algorithm: The Age of Killer Robots is Closer Than You Think," *Vox*, June 21, 2019, https://www.vox.com/2019/6/21/18691459/killer-robots-lethal-autonomous-weapons-ai-war.

2. Bonnie Docherty, "We're Running Out of Time to Stop Killer Robot Weapons," *The Guardian*, April 11, 2018, https://www.theguardian.com/commentisfree/2018/apr/11/killer-robot-weapons-autonomous-ai-warfare-un.

3. David Rivers, "Killer Robots Poised for 'MASS PRODUCTION' as Campaigners Urge AI to be Made ILLEGAL," *Daily Star*, August 21, 2018, https://www.dailystar.co.uk/news/latest-news/robots-artificial-intelligence-killer-war-16879356.

4. "'KILLER ROBOTS' WILL START SLAUGHTERING PEOPLE IF THEY'RE NOT BANNED SOON, AI EXPERT WARNS," *The Independent*, November 20, 2017, https://www.independent.co.uk/life-style/gadgets-and-tech/news/killer-robots-ban-autonomous-weapons-toby-walsh-ai-artificial-intelligence-un-amandeep-gill-a8065216.html.

5. "2018 Group of Governmental Experts on Lethal Autonomous Weapons Systems (LAWS)," United Nations Office at Geneva (website), accessed September 23, 2019, https://www.unog.ch/80256EE600585943/(httpPages)/7C335E71DFCB29D1C1258 243003E8724. "Report of the 2018 Session of the Group of Governmental Experts on Emerging Technologies in the Area of Lethal Autonomous Weapons Systems," October 23, 2018, https://undocs.org/en/CCW/GGE.1/2018/3.

6. "Country Views on Killer Robots," Campaign to Stop Killer Robots, November 22, 2018, https://www.stopkillerrobots.org/wp-content/uploads/2018/11/KRC_CountryViews-22Nov2018.pdf.

7. António Guterres, "Remarks at Web Summit," United Nations Secretary-General, November 5, 2018, https://www.un.org/sg/en/content/sg/speeches/2018-11-05/remarks-web-summit.

8. Samuel Gibbs, "Musk, Wozniak and Hawking Urge Ban on Warfare AI and Autonomous Weapons," *The Guardian*, July 27, 2015, https://www.theguardian.com/technology/2015/jul/27/musk-wozniak-hawking-ban-ai-autonomous-weapons.

9. "Lethal Autonomous Weapons Pledge," Future of Life Institute (website), accessed September 23, 2019, https://futureoflife.org/lethal-autonomous-weapons-pledge/.

10. Campaign to Stop Killer Robots (website), accessed September 23, 2019, https://www.stopkillerrobots.org/.

11. Stop Autonomous Weapons, "Slaughterbots," YouTube video, 7:47, November 12, 2017, https://youtu.be/9CO6M2HsoIA.

12. Michael Martinez, John Newsome, and Rene Marsh, "Handgun-Firing Drone Appears Legal in Video, but FAA, Police Probe Further," *CNN*, July 21, 2015, https://www.cnn.com/2015/07/21/us/gun-drone-connecticut/index.html.

13. Paul Scharre, "Why You Shouldn't Fear 'Slaughterbots,'" *IEEE Spectrum*, December 22, 2017, https://spectrum.ieee.org/automaton/robotics/military-robots/why-you-shouldnt-fear-slaughterbots.

14. Scharre, "Why You Shouldn't Fear 'Slaughterbots.'"

15. Scharre, "Why You Shouldn't Fear 'Slaughterbots.'"

1. THE NECESSITY OF LETHAL AI

1. Laura Hillenbrand, *Unbroken: A World War Story of Survival, Resilience, and Redemption* (New York: Random House, 2010).

2. "Holmes's skills: cryptography" https://monpinillos.wordpress.com/2008/06/02/holmess-skills-cryptography/

3. The Tor project website, https://www.torproject.org/

4. Nick Bilton, *American Kingpin: The Epic Hunt for the Criminal Mastermind Behind the Silk Road.* (New York: Penguin, 2017).

5. John Loeffler "How Peter Shor's Algorithm Dooms RSA Encryption to Failure" Interesting Engineering, May 2, 2019. https://interestingengineering.com/how-peter-shors-algorithm-dooms-rsa-encryption-to-failure

6. "Quantum Cryptology, Explained" Quantum Xchange. https://quantumxc.com/quantum-cryptography-explained/

7. Daniel Ogden and Robert J. Marks, "Daniel Ogden on Technology and National Security," July 25, 2019, on Mind Matters News (website), produced by The Walter Bradley Center for Natural and Artificial Intelligence, podcast, 43:38, https://mindmatters.ai/podcast/ep40/.

8. *The Man in the High Castle*, 2015- , Amazon Studios, https://www.imdb.com/title/tt1740299/.

9. Robert J. Marks II, ed., *Ed & Ray Hersman in WWII*, revised October 1, 2015, http://marksmannet.com/RobertMarks/ArticlesAndEssays/EdRayHersmanWWII.pdf.

10. Philip Jenkins, "Back to Hiroshima: Why Dropping the Bomb Saved Ten Million Lives, *ABC Religion & Ethics*, May 19, 2016, http://www.abc.net.au/religion/articles/2016/05/19/4465414.htm.

11. Jenkins, "Back to Hiroshima." Op. cit.

12. Tom Simonite, "Despite Pledging Openness, Companies Rush to Patent AI Tech," *Wired*, July 31, 2018, https://www.wired.com/story/despite-pledging-openness-companies-rush-to-patent-ai-tech/.

13. Rachel Sandler, "Peter Thiel Says CIA Should Investigate Google For Being 'Treasonous'" *Forbes*, July 15, 2019, https://www.forbes.com/sites/rachelsandler/2019/07/15/peter-thiel-says-cia-should-investigate-google-for-being-treasonous/#19e21c43521d.

14. Erik Sherman, "One in Five US Companies Say China Has Stolen Their Intellectual Property," *Fortune*, March 1, 2019, https://fortune.com/2019/03/01/china-ip-theft/.

15. Daniel Ogden and Robert Marks, "Daniel Ogden on Technology." Op. cit.

16. Sarah O'Meara, "Will China Lead the World in AI by 2030?" *Nature* 572 (August 21, 2019): 427-428, https://www.nature.com/articles/d41586-019-02360-7.

17. James Vincent, "Putin Says the Nation that Leads in AI 'Will be the Ruler of the World,'" The Verge (website), September 4, 2017, https://www.theverge.com/2017/9/4/16251226/russia-ai-putin-rule-the-world.

18. Samuel Bendett, "Putin Drops Hints about Upcoming National AI Strategy" Defense One (website), May 30, 2019, https://www.defenseone.com/ideas/2019/05/putin-drops-hints-about-upcoming-national-ai-strategy/157365/.

19. "Iran Ranks 8th for Top Papers in AI," *Tehran Times*, September 2, 2019, https://www.tehrantimes.com/news/439876/Iran-ranks-8th-for-top-papers-in-AI.

20. Dana Miller, "Iran Is Building an AI Supercomputer With or Without US Processors," *Interesting Engineering*, August 24, 2019, https://interestingengineering.com/iran-is-building-an-ai-supercomputer-with-or-without-us-processors.

21. Nick Anderson and Susan Svrluga "Universities Worry about Potential Loss of Chinese Students," *Washington Post*, June 3, 2019, "https://beta.washingtonpost.com/local/education/universities-worry-about-potential-loss-of-chinese-students/2019/06/03/567044ea-861b-11e9-98c1-e945ae5db8fb_story.html.

2. AI's Capabilities and Limits

1. See e.g. William Hsu and Joann G. Elmore. "Shining Light Into the Black Box of Machine Learning." JNCI: Journal of the National Cancer Institute (2019).

2. Kendra Cherry, "How Many Neurons Are in the Brain?" Very Well Mind, June 11, 2019. https://www.verywellmind.com/how-many-neurons-are-in-the-brain-2794889

3. Russell Reed and Robert J. Marks II, *Neural Smithing: Supervised Learning in Feedforward Artificial Neural Networks* (Cambridge, MA: MIT Press, 1999).

4. "Dragonfly," Wikimedia Foundation, last modified September 19, 2019, https://en.wikipedia.org/wiki/Dragonfly.

5. The word *convolution* is a mistitle. The mathematical operation performed is not convolution. It is *correlation*. The AI should be called the *deep correlation neural network*. See Robert J. Marks II, *Handbook of Fourier Analysis and Its Applications* (New York: Oxford University Press, 2009), chapter 2.

6. Bernard Widrow, "Science in Action," YouTube Video, 24:36, July 29, 2013, https://youtu.be/IEFRtz68m-8.

7. Segway website, https://www.segway.com/.

8. A. Khotanzad, R. Afkhami-Rohani, Lu Tsun-Liang, A. Abaye, M. Davis, D.J. Maratukulam, "ANNSTLF--a Neural-Network-Based Electric Load Forecasting System," *IEEE Transactions on Neural Networks* 8, no. 4 (July 1997): 835–846.

9. Michael Dunn, "Fair Isaac to Acquire HNC Software in $810 Million Stock Deal," *The Street*, April 29, 2002, https://www.thestreet.com/story/10019829/1/fair-isaac-to-acquire-hnc-software-in-810-million-stock-deal.html.

10. David B. Fogel, "Using Evolutionary Programing to Create Neural Networks that are Capable of Playing Tic-tac-toe," in *IEEE Proceedings of the 1993 International Joint Conference on Neural Networks* (March 1993): 875–880.

11. Kumar Chellapilla and David B. Fogel, "Evolving an expert checkers playing program without using human expertise." *IEEE Transactions on Evolutionary Computation* 5, no. 4 (September 2001): 422–428, doi: 10.1109/4235.942536. See also David B. Fogel, *Blondie 24: Playing at the Edge of AI* (San Francisco: Morgan Kaufmann, 2001).

12. Mark Robert Anderson, "Twenty Years on from Deep Blue vs Kasparov," The Conversation (website), May 11, 2017, http://theconversation.com/twenty-years-on-from-deep-blue-vs-kasparov-how-a-chess-match-started-the-big-data-revolution-76882.

13. Roger Penrose, *The Emperor's New Mind* (Oxford, UK: Oxford University Press, 2001); and Roger Penrose, *Shadows of the Mind: A Search for the Missing Science of Consciousness* (Oxford, UK: Oxford University Press, 1996).

14. Gregory Chirikjian, "Help Wanted for the Cognitive Era," JHU Robotics Center, Summer 2017, 44.

15. Satya Nadella, *Hit Refresh: The Quest to Rediscover Microsoft's Soul and Imagine a Better Future for Everyone* (New York: HarperBusiness, 2017), Kindle.

16. David Gelernter, "How Hard Is Chess?" *Time*, June 24, 2001, http://content.time.com/time/magazine/article/0,9171,137690,00.html.

17. Gary Smith, *The AI Delusion* (New York: Oxford University Press, 2018).

18. Natalia Wojcik "IBM's Watson 'Is a Joke,' Says Social Capital CEO Palihapitiya," CNBC, May 9 2017, https://www.cnbc.com/2017/05/08/ibms-watson-is-a-joke-says-social-capital-ceo-palihapitiya.html.

19. William A. Dembski and Robert J. Marks II, "Conservation of Information in Search: Measuring the Cost of Success," *IEEE Transactions on Systems, Man, and Cybernetics, Part A: Systems and Humans* 39, no. 5 (2009): 1051-1061.

20. Quotations by Roger Penrose, https://www-history.mcs.st-andrews.ac.uk/Quotations/Penrose.html.

21. Detection prior to classification can mitigate this. The point is that AI has no common sense when it comes to classification.

22. Peter Haas, "The Real Reason to be Afraid of Artificial Intelligence" TEDx talk, 12:37, December 15, 2017, https://youtu.be/TRzBk_KuIaM.

23. Su, Jiawei, Danilo Vasconcellos Vargas, and Kouichi Sakurai, "One Pixel Attack for Fooling Deep Neural Networks," *IEEE Transactions on Evolutionary Computation* (2019).

24. George Gilder, personal communication.

25. D. C. Park, M. El-Sharkawi, and R. J. Marks II, "An Adaptively Trained Neural Network," *IEEE Transactions on Neural Networks* 2 (1991): 334-345. R. J. Marks II, J.F. Walkup, and M. O. Hagler, "A Sampling Theorem for Space-Variant Systems," *Journal of the Optical Society of America* 66 (1976): 918-921.

26. Gwyn D'Mello, "Hong Kong Protestors Are Using Pocket Lasers To Fool Govt's Facial Recognition Cameras," *India Times*, August 13, 2019, https://www.indiatimes.com/technology/news/hong-kong-protestors-are-using-pocket-lasers-to-fool-govt-s-facial-recognition-cameras-373480.html.

27. Giedrė Vaičiulaitytė, "Artist Invents Jewelry That Will Make Your Face Unrecognizable With Facial Recognition Software," Bored Panda (website), accessed September 23, 2019, https://www.boredpanda.com/face-recognition-algorithms-incognito-mask-jewelry-ewa-nowak/?utm_source=google&utm_medium=organic&utm_campaign=organic.

28. Paul Scharre, *Army of None: Autonomous Weapons and the Future of War* (New York: WW Norton & Company, 2018), chapter 8.

29. Sam Shead, "Alphabet's DeepMind Losses Soared To $570 Million In 2018," *Forbes*, August 7, 2019, https://www.forbes.com/sites/samshead/2019/08/07/deepmind-losses-soared-to-570-million-in-2018/#1e8727335043.

30. "AI Puts Army's Electronic Warfare Missions in Focus" Meri Talk (website), March 29, 2019, https://www.meritalk.com/articles/ai-puts-armys-electronic-warfare-missions-in-focus/.

31. Sarvin Rezayat, Christopher Kappelmann, Zachary Hays, Lucilia Lamers, Matthew Fellows, Charles Baylis, Ed Viveiros, Abigail Hedden, John Penn, and Robert J. Marks, "Real-Time Amplier Load-Impedance Optimization for Adaptive Radar Transmitters Using a Nonlinear Tunable Varactor Matching Network," *IEEE Transactions on Aerospace and Electronic Systems* 55, no. 1 (2019): 160-169.

3. The Morality of Lethal AI

1. Cheryl Pellerin, "Project Maven to Deploy Computer Algorithms to War Zone by Year's End," US Department of Defense, July 21, 2017, https://www.defense.gov/Newsroom/News/Article/Article/1254719/project-maven-to-deploy-computer-algorithms-to-war-zone-by-years-end/.

2. "Amid Pressure from Employees, Google Drops Pentagon's Project Maven Account," *PBS News Hour Weekend*, June 3, 2018, https://www.pbs.org/newshour/show/amid-pressure-from-employees-google-drops-pentagons-project-maven-account.

3. Makena Kelly, "Microsoft Secures $480 Million HoloLens Contract from US Army," *The Verge*, November 28, 2018, https://www.theverge.com/2018/11/28/18116939/microsoft-army-hololens-480-million-contract-magic-leap.

4. Robert J. Marks, "Random Thoughts on the Passing Scene," Mind Matters News (website), August 1, 2019, https://mindmatters.ai/2019/08/random-thoughts-on-the-passing-scene-how-to-spell-gud/.

5. Adam Rosenberg, "Microsoft Workers Push Back Against Using HoloLens for US Military Training," *Mashable*, February 23, 2019, https://mashable.com/article/microsoft-military-contract-ivas-protest-hololens/#x0qNdZSYX8qy.

6. Alexander Moseley, "Just War Theory," *Internet Encyclopedia of Philosophy*, accessed September 23, 2019, https://www.iep.utm.edu/justwar/.

7. Phillipa Foot, "The Problem of Abortion and the Doctrine of the Double Effect," in *Applied Ethics: Critical Concepts in Philosophy*, eds. Ruth Chadwick and Doris Schroeder (Abingdon-on-Thames, UK: Routledge, 2001), 187.

8. George Washington, "First Annual Address, to Both Houses of Congress," January 8, 1790, https://www.mountvernon.org/library/digitalhistory/quotes/article/to-be-prepared-for-war-is-one-of-the-most-effectual-means-of-preserving-peace/.

9. "Cold War: A Brief History. Reagan's Star Wars" Atomic Archive (website), accessed September 23, 2019, http://www.atomicarchive.com/History/coldwar/page20.shtml.10. Nicholas Daniloff, review of *Reagan at Reykjavik: Forty-Eight Hours That Ended the Cold War* , by Kenneth Adelman, *Journal of Cold War Studies* 17, no. 2 (Spring 2015):145-148.

4. THE CHALLENGE OF IMPLEMENTING LETHAL AI

1. Paul Scharre, *Army of None.*

2. Eric Bonabeau, Marco Dorigo, Guy Théraulaz, and Guy Theraulaz, *Swarm Intelligence: from Natural to Artificial Systems* (New York: Oxford University Press, 1999).

3. Scharre, *Army of None*, chapter 12.

4. Sharif H. Calfee, "Autonomous Agent-Based Simulation of an AEGIS Cruiser Combat Information Center Performing Battle Group Air Defense Commander Operations," (thesis, Naval Postgraduate School, 2003).

5. Scharre, *Army of None*, introduction.

6. "Boeing X-45," Wikimedia Foundation,last modified July 3, 2019, 03:53, https://en.wikipedia.org/wiki/Boeing_X-45.

7. "USA and Israel in Crisis over China Harpy Deal," *Flight Global*, January 4, 2005, https://www.flightglobal.com/news/articles/usa-and-israel-in-crisis-over-china-harpy-deal-191940/.

8. Kirstie McCrum, "US Army Building 'Intelligent Drones' to Map Inside of Buildings During Times of War," *Mirror*, September 7, 2016, https://www.mirror.co.uk/news/world-news/army-building-intelligent-drones-map-8784030.

9. "Autonomous Underwater Vehicle," Wikimedia Foundation, last modified September 14, 2019, 03:46, https://en.wikipedia.org/wiki/Autonomous_underwater_vehicle.

10. This is law #1 of Asimov's *Three Laws of Robotics*. See "Three Laws of Robotics," Wikimedia Foundation, last modified September 22, 2019, 00:53, https://en.wikipedia.org/wiki/Three_Laws_of_Robotics.

11. Aaron Mamiit, "Rain or Snow? Rock or Plastic Bag? Google Driverless Car Can't Tell," *Tech Times*, September 2, 2014, https://www.techtimes.com/articles/14625/20140902/rain-or-snow-rock-or-plastic-bag-google-driverless-car-cant-tell.htm. Pedro Domingos, *The Master Algorithm: How the Quest for the Ultimate Learning Machine Will Remake Our World* (New York: Basic Books, 2015).

12. Seth Lloyd, "Measures of Complexity: a Nonexhaustive List," *IEEE Control Systems Magazine* 21, no. 4 (2001): 7-8.

13. Yu-Chi Ho, Qian-Chuan Zhao, and David L. Pepyne, "The No Free Lunch Theorems: Complexity and Security," *IEEE Transactions on Automatic Control* 48, no. 5 (2003): 783-793.

14. Winston Ewert, Robert J. Marks, Benjamin B. Thompson, and Albert Yu, "Evolutionary Inversion of Swarm Emergence Using Disjunctive Combs Control," *IEEE Transactions on Systems, Man, and Cybernetics: Systems* 43, no. 5 (2013): 1063-1076.

15. "Harpoon (missile)," Wikimedia Foundation, last modified September 20, 2019, 05:06, https://en.wikipedia.org/wiki/Harpoon_(missile).

16. "AGM/RGM/UGM-84 Harpoon Missile," Boeing (website), https://www.boeing.com/history/products/agm-84d-harpoon-missile.page.

17. Henry Gordon Rice, "Classes of Recursively Enumerable Sets and Their Decision Problems," *Transactions of the American Mathematical Society* 74, no. 2 (1953): 358-366.

18. Yu-Chi Ho, et al., "The No Free Lunch Theorems."

19. "Military Technology and AI," on Mind Matters News (website), produced by The Walter Bradley Center for Natural and Artificial Intelligence, podcast, 23:02, January 31, 2019, https://mindmatters.ai/podcast/military-technology-and-ai/.

20. "Iron Dome and Skyhunter Systems," Raytheon (website), https://www.raytheon.com/capabilities/products/irondome.

21. Scharre, "Why You Shouldn't Fear 'Slaughterbots."

22. Winston Ewert, Robert J. Marks II, Benjamin B. Thompson, and Albert Yu, "Evolutionary Inversion of Swarm Emergence Using Disjunctive Combs Control," *IEEE Transactions on Systems, Man & Cybernetics: Systems* 43, no. 5 (September 2013): 1063-1076.

Also, Jon Roach, Winston Ewert, Robert J. Marks II, and Benjamin B. Thompson, "Unexpected Emergent Behaviors From Elementary Swarms," in *Proceedings of the 2013 IEEE 45th Southeastern Symposium on Systems Theory (SSST)*, Baylor University (March 11, 2013) 41 - 50.

23. John Arquilla and David Ronfeldt, *Swarming and the Future of Conflict* (Arlington, VA: Rand Corporation, 2000).

24. Paul Scharre, "Why You Shouldn't Fear 'Slaughterbots."

5. LIVING IN A WORLD WITH LETHAL AI

1. Thanassis Cambanis, "The Logic of Assad's Brutality," *The Atlantic*, April 8, 2018, https://www.theatlantic.com/international/archive/2018/04/syria-chemical-weapons-assad-trump/557483/.

INDEX

www.ingramcontent.com/pod-product-compliance
Lightning Source LLC
Chambersburg PA
CBHW022132280326
41933CB00007B/652